Markov Models

Introduction to Markov Chains, Hidden Markov Models and Bayesian networks

Joshua Chapmann

MARKOV MODELS

Table of Contents

Chapter 1 – Introduction to Markov Models 6

Chapter 2 – About the Series 8

Chapter 3 – Foundations of Markov Processes 11

Chapter 4 – Markov chains 14

Chapter 5 – Case Study: Google PageRank 27

Chapter 6 – Hidden Markov Models 31

Chapter 7 – Bayesian networks 37

Chapter 8 – Inference tasks 40

Conclusion 48

© Copyright 2017 by Joshua Chapmann - All rights reserved.

This document is geared towards providing exact and reliable information in regards to the topic and issue covered. The publication is sold with the idea that the publisher is not required to render accounting, officially permitted, or otherwise, qualified services. If advice is necessary, legal or professional, a practiced individual in the profession should be ordered.

Legal Notice: From a Declaration of Principles which was accepted and approved equally by a Committee of the American Bar Association and a Committee of Publishers and Associations. In no way is it legal to reproduce, duplicate, or transmit any part of this document in either electronic means or in printed format. Recording of this publication is strictly prohibited and any storage of this document is not allowed unless with written permission from the publisher. All rights reserved.

Disclaimer Notice: The information provided herein is stated to be truthful and consistent, in that any liability, in terms of inattention or otherwise, by any usage or abuse of any policies, processes, or directions contained within is the solitary and utter responsibility of the recipient reader. Under no circumstances will any legal responsibility or blame be held against the publisher for any reparation, damages, or monetary loss due to the information herein, either directly or indirectly. Respective authors own all copyrights not held by the publisher. The information herein is offered for informational purposes solely, and is universal as so. The presentation of the information is without contract or any type of guarantee assurance.

Chapter 1 – Introduction to Markov Models

Markov models are a powerful predictive modelling technique developed by one of probability theory's founding fathers, the Russian mathematician Andrey Markov. These are used to model stochastic systems, i.e.

systems with randomly changing outcomes. They are built around the concept of "memoryless" modelling, which states that the outcome of a problem depends only on the current state of the system; historical data is has no effect on the next outcome.

This construction may sound straightforward and counter-intuitive. After all, if you have historical data available why not use it to develop more complete and well-informed models? Surely, it would lead to more accurate predictions. However, when modelling time-series data where previous results are of limited relevance, a memoryless model delivers vast performance advantages.

By considering only one state, algorithms become highly scalable, stable, fast and above-all-else extremely versatile.

Speech recognition is a perfect application - it is a continuous stream of data where the past results are (largely) irrelevant. What was translated three sentences ago has no impact on the translation of the next word – it can be ignored. In fact, the majority of modern speech recognition algorithms are built using Markov Models.

Throughout the course of this book we will explore why a memoryless predictive model can be so advantageous to the modern tech industry at large. We will take a look at fundamental mathematics and high-level concepts alike, extending our understanding of the subject beyond a simple model. In particular, we will discuss:

- Foundations of Markov Models
- Markov Chains
- Case Study: Google PageRank
- Hidden Markov Models
- Bayesian Networks
- Inference Tasks

Chapter 2 – About the Series

"Markov Models: Introduction to Markov Chains, Hidden Markov Models and Bayesian networks" is the third instalment of the book series **Advanced Data Analytics**, carefully developed by myself and a team of software-loving engineers. This series will provide you with an

introduction into the world of modern data analytics. The material covered is roughly comparable to a 1-semester introductory course in Artificial Intelligence.

Throughout the series I only assume a high-school level knowledge in mathematics and statistics and absolutely no previous exposure to computing or coding. Whenever we come across a new topic, concept or formula I ensure to cover all the required material beforehand, maximizing and facilitating your learning process.

However, my explanations can only go *so far*. Please understand this series will challenge and push your understanding of the modern tech world, revealing many applications of computer and algorithms you never thought possible. Especially in later books, we will dive into very technical topics at the forefront of research. To follow along and keep up with the material, I need you to be

committed and **passionate** about the topics we will cover.

The material across the three-book series was designed to complement and work together, therefore I recommend working across all 3 books for the most complete learning experience. I would highly appreciate any feedback on the current publications and suggestions for future topics – please leave these in Amazon's official review section.

Editor's Note: The previous instalments of the Data Analytics Book series gave readers a general introduction to the world of Machine Learning algorithms and Neural Networks. These books are by no means required pre-reading, but provide an excellent context for the material discussed in later chapters and may simplify the way you view and absorb the material later discussed. You can find all books in the series on my official Amazon Author page.

If you are ready, let's now dive into the world of Markov Models!

Chapter 3 – Foundations of Markov Models

In this chapter we will discuss fundamental definitions and the quintessential Markov Property

Definitions

The first two definitions we must consider are: **states** and **intelligent agents.** These are present in every modelling scenario, whether real or fictional. A state contains all the information required to predict the effect of an **action** and to determine if it is a goal state (what we wish to achieve). Agents are autonomous entities that observe a world using **sensors**.

For example, imagine yourself throwing a pair of dice. The

state are two dice rolling on the table, the action is throwing the dice and you are the agent. You are always able to see the dice, which makes this a **fully observable** environment.

Even though this event is fully observable and you know the past dice results, you cannot know what the outcome of this throw will be. In other words, the next state is *not* determined by the previous state, but depends only on the agent's action. This is known as a **stochastic** (or in other words **random**) system and represents the basis for the Markov Property.

The Markov Property

In the early 1900s, the Russian mathematician **Andrey Markov,** who specialized in stochastic processes, explains that a process has the Markov property if future states are *only* dependent on the current state and not on any previous states. This means that past states have no effects on future outcomes.

Using standard statistics notation, we can state that a state at time *t* can be determined *only* by the state at *t-1*, meaning, and that states at *t-2*, *t-3*, etc. are irrelevant. This is illustrated by the formula below.

$$P(X_t \mid X_{0:t-1}) = P(X_t \mid X_{t-1})$$

As the probability distribution is not dependent on the

amount of time that has elapsed or states beyond t-1, the Markov Property can be described as **memoryless**.

The construction of Markov models was without a doubt impressive at the time; it was the first highly complex calculations by Mr. Markov himself, as we will see in later chapters. However, it is only with modern increases in computational power that we can implement these models on a wide scale for **predictive modelling** and **probabilistic**

forecasting. Today, you will find widespread use of Markov models throughout modern computerized world such as the PageRank, which is an algorithm developed by Google to rank web search results. You will also find the use of Markov models in your smartphone. When typing, the predictive text suggestions are generated using Markov models. There are numerous other real-life applications of Markov Models and we will come across a variety throughout the course of this book.

Chapter 4 – Markov chains

The weak law of large numbers

Imagine that you have a bag of marbles with two colors, red and blue. The ratio of blue marbles to red marbles is 3 to 2. It follows that if you have 100 marbles in total, 60 will be blue and 40 will be red. If you draw a marble at random with replacement, as the number of draws approaches infinity, you will find that the ratio of drawn marbles approaches 3 to 2. This is known as the **weak law of large numbers**.

It is important to note that in this example you are drawing the marbles with replacement, which means each event is **independent** of each other.

It does not matter how many blue/red marbles you drew previously; in the next draw there will be exactly 60 blue marbles and 40 red marbles to choose from. Therefore, the odds of picking a blue/red in the next draw are not affected by previous draws.

Before Markov, mathematicians believed the only correct way of modelling this draw in accordance with the weak law of large numbers was through a single independent event.

Markov did not agree with this; he believed that even dependent stochastic events follow the weak law of large numbers, if linked correctly. To prove his point, Markov developed the Markov Chain.

Deriving Markov Chain

To prove that dependent stochastic events abide the law of large numbers, Markov used two bags filled with marbles. In the first bag named **S_0** he placed 50 blue and 50 red marbles, creating a 1 to 1 distribution of each. In the second bag named **S_1**, he placed 65 blue and 35 red marbles, producing a 13 to 7 distribution of blue to red marbles. His model worked in the following way:

You begin drawing from the bag with 1:1 distribution. If you get a red marble, you put it back and redraw from the same bag. If you pick a blue, you will put it back in the same bag, but your next draw will be from bag 2. If you draw a blue marble from bag 2, you will put it back and draw again until you obtain a red marble. When this happens, you put it back and return to bag 1.

This model is a **Markov Chain** and can therefore be represented using the **state transition diagram** shown on the next page.

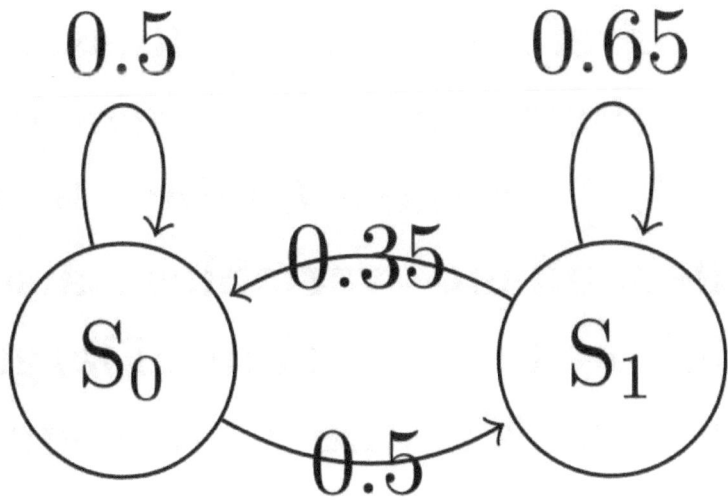

In this system each draw is a dependent event, i.e. you only draw from bag 2 if you found a blue marble in bag 1. However, as the number of draws approaches infinity the results from both bags agree with the large law of weak numbers.

Therefore, the Russian Mathematician proved, using Markov Chains, that dependent variables also follow the weak law of large numbers as the event count approaches infinity,

provided each state is reachable.

Key Features of Markov Chains

As we can see from the state transition diagram, there are several key features that define a Markov chain. First, we have a set of states, $S = \{s_1, s_2, ..., s_r\}$ which represent all the possible outcomes of an event. A process starts in one of the states and moves from one state to another.

In the example above, say we start in S_0, then you can either stay in S_0, by 'moving' to S_0 again or you can move to S_1. Likewise, if you are in state S_1, you can stay in S_1 by looping back on yourself, which in this case is likely because S_1 has a 0.65 probability of looping back on itself and only a 0.35 probability of moving back to S_0.

This can be generalized to any state and we can say: the

probability of moving from state S_i to S_j is controlled by the term P_{ij}, called the **transition probability**. As previously mentioned, states can loop back on themselves, which gives a probability P_{ii}.

As Markov Models increase in size, a **Transition Matrix** is used to store all the transition probabilities between states.

A very common real-life application of this principle is weather prediction. This is because global weather patterns are often too complex to model using mathematical patterns and a simplified stochastic approach is common. In this technique, only the current state is assessed (e.g. current wind speed, rainfall, cloud pattern formation and movement, etc.) and previous patterns are ignored, as building mathematical patterns is far too complex. Then, based on the current state transition probabilities of moving between state is assigned.

In the simplified model, I have created a model with 3 states 'Nice', 'Rainy' and 'Snowy' and reported its transition matrix below. Here, the column represents the current state and the row represents the future state; the transition probability is the corresponding number. Using this matrix, we can immediately see that there will never be two nice days in a row.

$$P = \begin{array}{c} \\ R \\ N \\ S \end{array} \begin{array}{ccc} R & N & S \\ \begin{pmatrix} 1/2 & 1/4 & 1/4 \\ 1/2 & 0 & 1/2 \\ 1/4 & 1/4 & 1/2 \end{pmatrix} \end{array}$$

Say we know for a fact, that the weather today is raining. We can use this using a **state distribution matrix** where S_0 = [1 0 0]. State distribution matrices are used to represent the probability of each state occurring for a given event. In this particular case the 1.00 represents 100% probability.

To predict future states, we must simply multiply the state

distribution and state transition matrix. For example, given that we know the current state distribution matrix S_0 = [1 0 0], let's predict the state distribution state in two days S_2, using our transition matrix.

First, we must multiply S_0 and the transition matrix, to obtain the state distribution matrix in 1 day, S_1:

$$S_1 = \begin{bmatrix} 1 & 0 & 0 \end{bmatrix} \begin{bmatrix} \frac{1}{2} & \frac{1}{4} & \frac{1}{4} \\ \frac{1}{2} & 0 & \frac{1}{2} \\ \frac{1}{4} & \frac{1}{4} & \frac{1}{2} \end{bmatrix}$$

$$= \begin{bmatrix} 0.5 & 0.25 & 0.25 \end{bmatrix}$$

Using the same principle, we can find the 2nd day state distribution matrix S_2, by multiplying S_1 and the transition matrix as shown on the following page.

$$S_2 = \begin{bmatrix} 0.5 & 0.25 & 0.25 \end{bmatrix} \begin{bmatrix} \frac{1}{2} & \frac{1}{4} & \frac{1}{4} \\ \frac{1}{2} & 0 & \frac{1}{2} \\ \frac{1}{4} & \frac{1}{4} & \frac{1}{2} \end{bmatrix}$$

$$= \begin{bmatrix} (0.5 \times \frac{1}{2} + 0.25 \times \frac{1}{2} + 0.25 \times \frac{1}{4}) & (0.5 \times \frac{1}{4} + 0.25 \times 0 + 0.25 \times \frac{1}{4}) & (0.5 \times \frac{1}{4} + 0.25 \times \frac{1}{2} + 0.25 \times \frac{1}{2}) \end{bmatrix}$$

$$S_2 = \begin{bmatrix} 0.4375 & 0.1875 & 0.375 \end{bmatrix}$$

Using the Markov Chain, we see that the probability of rain in 2 days is 0.4375, assuming it rained on day 0 and my randomly-assigned transition state matrix.

Chapman-Kolmogrov Equation

The Chapman-Kolmogrov equation is a quick and convenient tool to quickly find the transition matrix for any future states after n number of events. For example, to find the probability matrix of rain after 2 days, simply take the transition matrix to the power of 2 as shown on the next page.

$$\begin{bmatrix} \frac{1}{2} & \frac{1}{4} & \frac{1}{4} \\ \frac{1}{2} & 0 & \frac{1}{2} \\ \frac{1}{4} & \frac{1}{4} & \frac{1}{2} \end{bmatrix}^2$$

$$= \begin{pmatrix} 0.4375 & 0.1875 & 0.375 \\ 0.375 & 0.25 & 0.375 \\ 0.375 & 0.1875 & 0.4375 \end{pmatrix}$$

As we saw in earlier examples, the top row represents the probability of rain. This matches the result obtained earlier through hand-calculations for each step. Furthermore, we get the entire distribution so we also know the probability of nice weather and snow in 2 days, for all possible current states.

The Chapman-Kolgorov equation will work for any step in the chain and can be represented using the formula below, here u is the current state distribution, $u^{(n)}$ is the state distribution in n days and P^n is the transition matrix to the power n.

$$u^{(n)} = uP^n$$

Note, in the previous example we did not need to worry

about the initial stat distribution u and simply assumed 100% rain in the present. I used this simplification for a first example, but a realistic scenario may be rain with a 10% chance of snow, which we represent [0.9 0 0.1].

Stationary Matrix

We can use the Chapman-Kolgorov equation to find the probability matrix for any step. For example, below you will find the distribution in 30 days.

$$= \begin{pmatrix} 0.40000... & 0.20000... & 0.40000... \\ 0.4 & 0.2 & 0.4 \\ 0.40000... & 0.20000... & 0.40000... \end{pmatrix}$$

What you notice as you reach a higher and higher number of steps, is that the value in the distribution converges. We notice that the values for rain and snow reach 0.4, while nice weather reaches 0.2. For step 30 you would need to round a couple decimal places, however, as you approach infinity, these values will all converge and eventually end up as 0.4 and 0.2 without the need for rounding. This is what is known as the **stationary matrix** and is a property that always exists for **regular Markov chains**. You will notice this is a reappearance of the weak law of large numbers we observed earlier in a bag of marbles, but applied to a state distribution diagram.

An effective way to check our stationary matrix has converged, is to multiply our convergence and transition matrix. If convergence has reached, the result should not change. This is shown below.

$$\begin{bmatrix} \frac{1}{2} & \frac{1}{4} & \frac{1}{4} \\ \frac{1}{2} & 0 & \frac{1}{2} \\ \frac{1}{4} & \frac{1}{4} & \frac{1}{2} \end{bmatrix} \times \begin{bmatrix} 0.4 & 0.2 & 0.4 \\ 0.4 & 0.2 & 0.4 \\ 0.4 & 0.2 & 0.4 \end{bmatrix}$$

$$= \begin{bmatrix} 0.4 & 0.2 & 0.4 \\ 0.4 & 0.2 & 0.4 \\ 0.4 & 0.2 & 0.4 \end{bmatrix}$$

From this result a further conclusion can be reached: For events far enough into the future the current state has no effect. This is sensible and matches the memoryless nature of Markov Chains.

Regular Markov chains

A Regular Markov Chain is one that has a **Regular Transition Matrix**. The definition of a regular transition matrix is that, when taken to a power n, the resulting matrix will only contain values greater than 0.

Looking back at the transition matrix for the earlier weather example, we may initially think it is not regular because the probability of rain two days in a row is zero, or $P(N_1|N_0) = 0$. However, we observed that when escalating the matrix to n>2, all values were >0. This was further proved by the stationary matrix.

Alternative Ways to Find the Stationary Matrix

Previously, we used the Chapman-Kolmogrov equation to find the stationary matrix and the numbers they converge to. However, this method can prove very tedious and labor intensive. Additionally, it can sometimes difficult to determine when convergence has been reached exactly.

However, we can also reach the stationary matrix without calculating individual steps. In fact, from the properties discussed earlier we know that if a converged stationary matrix at convergence is multiplied by the transition matrix, the answer should be unchanged. This principle can be expressed in the following manner:

$$\begin{bmatrix} R & N & S \end{bmatrix} \times \begin{bmatrix} \frac{1}{2} & \frac{1}{4} & \frac{1}{4} \\ \frac{1}{2} & 0 & \frac{1}{2} \\ \frac{1}{4} & \frac{1}{4} & \frac{1}{2} \end{bmatrix} = \begin{bmatrix} R & N & S \end{bmatrix}$$

We can use this to formulate a set of equations:

$$0.5R + 0.5N + 0.25S = R$$
$$0.25R + 0N + 0.25S = N$$
$$0.25R + 0.5N + 0.5S = S$$

We also know that the sum of the vector should be one, so we can add an additional equation:

$$R + N + S = 1$$

We can then use any method to solve this set of equations; I will use substitution. This gives us:

$$R = 1 - N - S$$
$$N = 1 - R - S$$
$$S = 1 - R - N$$

We can then substitute this

$$0.25(1 - N - S) + 0.25S = N$$
$$0.25 - 0.25N - 0.25S + 0.25S = N$$

into our original equations:

$$0.25 - 0.25N = N$$

S cancels out which gives us:

We then have to add 0.25N to both sides, which gives:

$$0.25 = 1.25N$$
$$N = 0.2$$

$$0.5(1 - 0.2 - S) + 0.5(0.2) + 0.25S = 1 - 0.2 - S$$
$$0.5 - 0.1 - 0.5S + 0.1 + 0.25S = 0.8 - S$$
$$0.5 - 0.25S = 0.8 - S$$
$$0.75S = 0.3$$
$$S = 0.4$$

We can then use equation 1:

Having retrieved both N and S, we then can easily find R

$$R + N + S = 1$$
$$R + 0.2 + 0.4 = 1$$
$$R + 0.6 = 1$$
$$R = 0.4$$

This gives us the stationary matrix:

$$[0.4 \quad 0.2 \quad 0.4]$$

This matrix shows that at any time and regardless of initial conditions, the probability of rain and snow are 0.4 and nice weather is 0.2. This tells us that if we are going on holiday to this place and we are staying 20 days, then we should expect 4 days of nice weather, 8 days snow and 8 days rain.

Now, we have explored Markov chains by explaining how they are implemented in a simple weather prediction model.

Chapter 5 – Case Study: Google PageRank

In this chapter, we will briefly discuss how Markov Models are implemented in the Google PageRank algorithm. This is one of Markov Model's most famous and intriguing application, therefore I find it very useful for students to consider.

When ordering the results of a Google search, you want results that have many instances of your search query. However, if this was the only requirement and a person would search the word "Disney", the website with the greatest repetitions of the word would rank 1st. Clearly, this is not a very effective method.

Google therefore adopted a different method, where

instead of checking the amounts of query repetition, it looks at numbers of **in-links** present in a webpage.

In-links are links around the internet pointing to that webpage. This is a better representation of quality and will help in the ranking process. However, one could rank very high by simply creating a large network of webpages pointing to the chosen website, thus increasing the number of in-links.

As a result, the PageRank algorithm is designed to weigh in-links from high-quality sites more heavily. High quality websites are those that have a large number of in-links by other high-quality websites. Clearly, this creates a looping definition and can appear confusing. But can be clarified by using the formula below:

$$PR(p) = \frac{1-d}{N} + d\sum_{i} \frac{PR(in_i)}{C(in_i)}$$

Where *PR(p)* is the PageRank of a given page *p*, *N* is the number of pages in body and In_i refers to the number of pages linking to the page *p*. $C(in_j)$, refers to the count of the total number of out-links on page in_i. Lastly, we have *d*, which refers to the damping factor. To define the damping factor, we must consider the **Random Surfer Model.**

Imagine a web surfer who starts at some random page and begins exploring. With probability *d*, the surfer clicks on one of the links on the page, and with probability *1 - d*, the surfer gets bored with the page and restarts a new surf session from a random page, anywhere on the web. From this, we understand the damping factor represents the probability that the surfer will click on a link on the page or start a new session. In most cases this damping factor is set taken as 0.85.

Furthermore, we can observe that the Random Surfer Model is in fact a Markov Chain. In fact, websites are modelled as states connected by a massive probability transition matrix.

More precisely, it is a non-reversible Markov Chain, because the surfer has a 15% probability of not clicking the link and starting from a new random page, and cannot return to the previous page.

The final ranking is calculated iteratively by having every page initially starting with *PR(p) = 1* and iterating the Markov Chain, updating ranks until they converge and the stationary matrix is reached. The ranking of a page *p* is then simply the probability that the random surfer will find himself of page *p* at any point in time.

Chapter 6 – Hidden Markov Models

So far in this book we have only discussed fully observable environments. Meaning, you can see all states at any one time. However, this is not always the case. Sometime a world is only partially observable and we have to make assumptions based on the limited information we are able to witness.

Rather than seeing a state, we have to try and determine what it is based on evidence. This is known as a **Hidden Markov model (HMM)**. These are typically used to describe probability distributions over a chain of observations. The general transition diagram for an HMM is shown below:

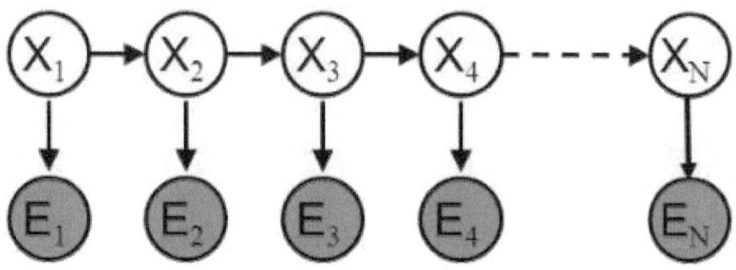

We would therefore define an HMM as a sequence of random variables, X_1, X_2, ..., X_N such that the probability distribution of X_N depends only on the evidence E_N of an associated Markov chain. We can say that and HMM is a

Markov chain, where not all the variables are visible.

This is clear because if you removed the E variables from the above diagram, you would be left a Markov chain. However, how do the evidence variables E_N help determining the value of X_N? In the HMM diagram you will notice that the arrows point from the state (X_N) to the evidence (E_N), meaning the state is dependent on the evidence.

This dependency means that for a given state, you should be able to observe the corresponding evidence. Hence, we can define the **emission probability** as shown below.

$$P(E_t \mid X_{0:t}, E_{0:t-1}) = P(E_t \mid X_t)$$

(Note t and N are used interchangeably but always denote the number of states in the system)

This is known as the **sensor model** (or observation model). If we know what the probability of certain evidence given a state is, then we can then use the evidence for the following time $t + 1$ to calculate X_{t+1}. However, it is not enough only having the sensor model, we also need to know the probability of transitioning between two states are, what we earlier defined as the probability transition model.

$$\mathbf{P}(\mathbf{X}_i \mid \mathbf{X}_{i-1})$$

Finally, to calculate a state X_τ we also need to know the initial state model.

$$\mathbf{P}(\mathbf{X}_0)$$

This leaves us with the

$$\mathbf{P}(\mathbf{X}_{0:t}, \mathbf{E}_{1:t}) = \mathbf{P}(\mathbf{X}_0) \prod_{i=1}^{t} \mathbf{P}(\mathbf{X}_i \mid \mathbf{X}_{i-1}) \mathbf{P}(\mathbf{E}_i \mid \mathbf{X}_i)$$

product as shown below.

At first glance, this formula may appear a little daunting, but upon closer inspection we realize it is remarkably similar to the formula for Markov Chains. In fact, the only difference is the addition of the probability of a piece of evidence at time t given the state X at the same time t. To explain HMM in more depth I will work through the **Umbrella World** example, one of the most common examples used to teach HMMs:

Imagine you are a security guard stationed at a secret underground station. You want to know if it is raining today, but your only interaction with the outside world occurs each morning when you see the director coming in with, or without, an umbrella. For each day then, E_T contains a single evidence variable U_T, which represents whether the director has brought an umbrella or not. Likewise, X_T also contains a single variable R_T, used to denote whether is raining or not. This is illustrated by the diagram on the following page.

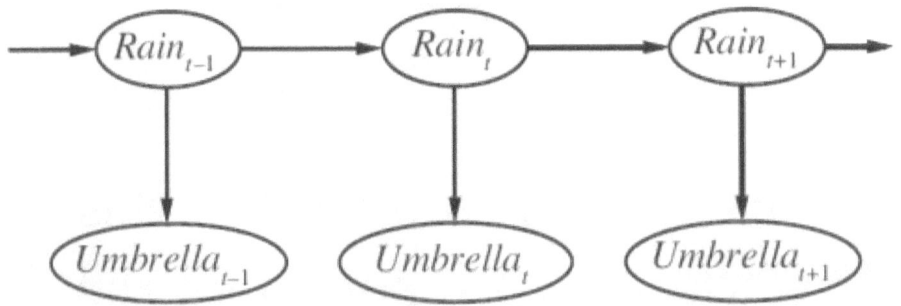

The transition matrix for this world is shown below and explains that if it rained yesterday, there is 0.7 chance it will be raining again today. If it was not raining yesterday, there is a 0.3 chance it will rain today.

R_t	R_{t-1}=no	R_{t-1}=yes
no	0.7	0.3
yes	0.3	0.7

The **sensor model** for this world is also reported on the following page. This states that if it rained today, there is a 0.9 probability that the director will bring an umbrella. In other words, there is a 0.1 probability that the director does not bring the umbrella even when it rains. It is not guaranteed that if there is no umbrella, it is not raining.

U_t	R_t=no	R_t=yes
no	0.8	0.1
yes	0.2	0.9

The only parameter missing from our formula is the initial state model, which we will assume to be 100% sunny. This is denoted by $X_0 = s$. Hence, we can now use the HMM formula to calculate the probability that it will be sunny at time $t = 1$.

Chapter 7 – Bayesian networks

In this chapter we will discuss Bayesian networks (BNs), which are a more generalized approach to represent conditional independencies over HMMs.

In the last chapter, we will will use BNs to help us understand the various used in inference for HMMs. In essence, BNs are graphical models used for representing knowledge about an uncertain domain. In other words, they show conditional independencies, where each node in the graph illustrates a random variable, while the edges between nodes represent the probabilistic dependencies between the random variables.

A great example of a Bayesian network is the Monty hall problem and notoriously depicted in the blackjack statistics movie "21". Based on the TV gave show 'Let's make a deal', contestants are given the opportunity to choose between 3 doors. Behind 2 of the doors there is nothing, but behind the remaining door is a brand-new car. Now, this seems quite simple: 3 doors, 1/3 chance of winning.

However, after you select a door Monty (the Host) will show you a door with nothing behind it and ask if you would like to change your choice (you can pick any of the remainder 2). You might think you now have the same odds of winning if you switch, but by using a BN to model the problem, we see that this is not the case.

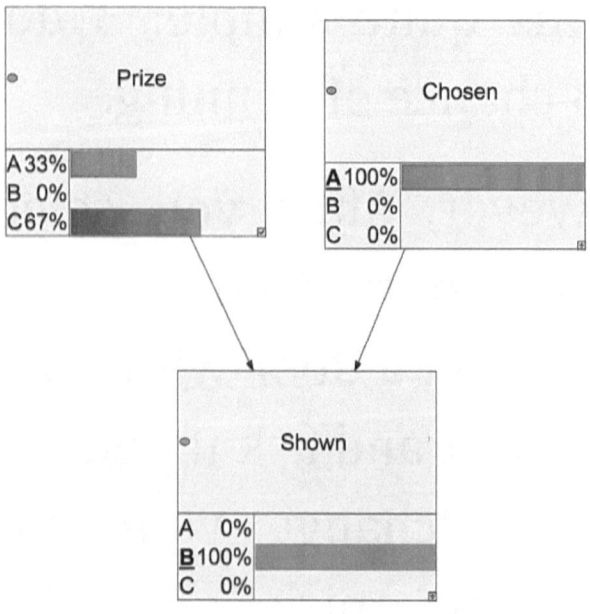

The diagram shows that the chosen value is A and Monty shows you door B. Now the probability that the prize is behind door B, is 0 because this is the door Monty showed you. However, the probability that the prize is behind door A, your original choice is only 1/3 but having switched door would have given you a probability of 2/3 of winning the prize. The reason is, if you select a door with nothing behind it on the first try and switch, you will always win. The probability that you pick a door with nothing behind it is 2/3 because out of the 3 doors, 2 of them have nothing behind them. This also explains why not switching gives you a 1/3 probability of winning, because you only have 1/3 doors containing the prize.

Chapter 8 – Inference tasks

Having briefly covered BNs, we can how move into inference and their role in Hidden Markov Models.

N^{th} order Markov Model

Firstly, we must revisit the definition of the Markov Property. In chapter 3, we defined this term as a stochastic state where the current state only depends only on the previous state. However, in many real-life stochastic events knowing the previous state is not sufficient and more information is required.

Therefore, a more accurate definition of the Markov Property would be: **the current state depends on only a finite number of previous states**. This is where the term **n^{th}-order**

Markov Process stems from, and it states that the probability of a given state is dependent on the *n* previous states.

$$P(X_t|X_{t-n}, ..., X_{t-1})$$

Forward Inference Algorithm

The first inference algorithm we will look at is the **Forward** inference algorithm. This algorithm is an example of what is known as **filtering**. The forward algorithm computes the probability of being in a state at a certain time t based on all previous evidence. That is, we are trying to compute the following.

$$P(X_t, e_{1:t})$$

We can then use this to get the formula below, where α_t

$$\alpha_t(X_t) = \sum_{X_{t-1}=1}^{m} P(X_t X_{t-1}, e_{1:t})$$

symbolizes the formula.

$$= \sum_{X_{t-1}} P(e_t|X_t, X_{t-1}, e_{1:t-1}) P(X_t|X_{t-1}, e_{1:t-1}) P(X_{t-1}, e_{1:t-1})$$

We then factor this out and get.

However, using the Markov assumption, we can simplify this expression quite a bit further:

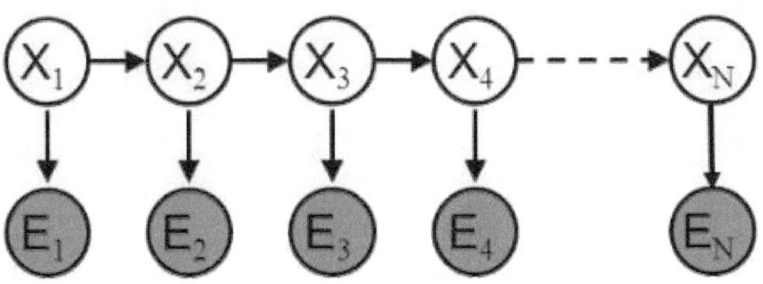

Here we can see that E_N (or e_t) is conditionally independent of all values except X_N (or X_t). This means that we can remove X_{t-1} and $e_{1:t-1}$ from the

$$= \sum_{X_{t-1}} P(e_t|X_t)P(X_t|X_{t-1},e_{1:t-1})P(X_{t-1},e_{1:t-1})$$

formula, which leaves us with:

We can continue simplifying the expression. Looking back at the diagram, we can see that given X_{t-1}, X_t is conditionally independent on all evidence up to *t-1*, so we

$$= \sum_{X_{t-1}} P(e_t|X_t)P(X_t|X_{t-1})P(X_{t-1},e_{1:t-1})$$

can remove that from the equation as well.

Now if we look at this formula now, it is not overly complicated if we separate it into pieces that we can relate to. The first part, is simply the emission probability, the probability of evidence (observation) given a certain hidden variable.

$$P(e_t|X_t)$$

The next term is simply the transition probability, the

probability of moving from one state to another.

$$P(X_t|X_{t-1})$$

In the last part we have simply the previous step of what we are trying to compute and we can denote it as $\alpha_{t-1}(X_{t-1})$.

$$P(X_{t-1}, e_{1:t-1})$$

This gives us the recursion

$$\alpha_t(X_t) = \sum_{X_{t-1}}^{m} P(e_t|X_t)P(X_t|X_{t-1})P(X_{t-1}, e_{1:t-1})$$

function as shown below.

This formula of course only applies for t = {2, ..., n}, because we cannot have a value less than one for the term *t–1*. Furthermore, as this is a recursive algorithm, and therefore we need to define a base case: we need our base case: that is $\alpha_1(X_1)$:

$$\alpha_1(X_1) = P(e_1, X_1) = P(X_1)P(X_1|e_1)$$

We will know $P(X_1)$ because it is our initial distribution and we also know the second part $P(X_1|e_1)$ because as mentioned before, it is just the emission probability. This is how the Forward inference algorithm can be applied to HMM.

Smoothing Inference Algorithm

The next technique we will consider is **smoothing** and more specifically, the **forward-backward** algorithm. For every hidden variable (X_t), the forward-backward algorithm computes the probability distribution given a sequence of evidence, this is known as the **posterior marginals**. It makes use of a principle called **dynamic programming**, which means taking a complex problem and breaking it into a collection of simpler problems. Dynamic programming is used in the forward-backward algorithm to get the posterior marginal distributions in two passes.

The first pass goes forward in time, as explained in the forward algorithm and the second pass goes backward in time. In other words, the forward-backward algorithm is trying to compute the following:

$$P(X_t, e_{1:n})$$

As explained, there are two steps to computing this algorithm. Step 1 is the forward algorithm, which we covered in the previous section. Step 2 is the **backward** algorithm, where the goal is to compute the following.

$$P(e_{t+1:n}|X_t)$$

To compute this, we will adopt an approach very similar to that we used when computing the forward algorithm, namely the

$$\beta_t(X_t) = \sum_X {}_{t+1=1}^m P(e_{t+1:n}, X_{t+1}|X_t$$

$$= \sum_{X_{t+1}} P(X_{t+2:n}|X_{t+1}, X_t, e_{t+1})P(e_{t+1}|x_{t+1}, X_t)P(X_{t+1}|X_t)$$

creation of a recursive algorithm.

Like in our example with the forward algorithm, we can use the Markov properties to simplify the expression by removing values where there is conditional independence.

$$= \sum_{X_{t+1}} P(X_{t+2:n}|X_{t+1})P(e_{t+1}|x_{t+1})P(X_{t+1}|X_t)$$

This leaves us with the following:

Again, like in the forward algorithm, we can split this

expression into parts. The first part is $\beta_{t+1}(X_{t+1})$. What remains then is the emission probability and the transition probability, both of which will be known. We can illustrate

$$\beta_t(X_t) = \sum_{X_{t+1}=1}^{m} \beta_{t+1}(X_{t+1})P(e_{t+1}|X_{t+1})P(X_{t+1}|X_t)$$

this by the following.

This is then valid for t = {1, 2, ..., n-1} because we cannot have $t + 1 > n$. Therefore, our base case for the recursive algorithm will be $\beta_{n-1}(X_{n-1})$, which is equal to 1.

$$\beta_{n-1}(X_{n-1}) = 1$$

Now that we have shown how to calculate both the forward and backward algorithm, it is time to combine the two, to

$$P(X_t|e_{1:n}) \propto P(X_t, e_{1:n}) = P(e_{t+1:n}|X_t, e_{1:t})P(X_t, e_{1:t})$$

create the forward-backward algorithm.

We can tell that $e_{1:t}$ is conditionally independent of $e_{t+1:n}$ given X_t, and therefore we can remove it from the equation to simplify it, much like with both the forward and backward algorithm. This leaves the following formula.

$$P(X_t|e_{1:n}) \propto P(X_t, e_{1:n}) = P(e_{t+1:n}|X_t)P(X_t, e_{1:t})$$

Now, if we split the formula (after the =), we can see that the first part is the backward algorithm and the second part is forward algorithm.

Most Likely Explanation

The last inference task we will look at is the **most likely explanation**, that is what is the most likely path in an

HMM? The algorithm which will be used to calculate this, is known as the **Viterbi** algorithm. Viterbi is a dynamic programming algorithm used for finding the most likely sequence of hidden states. What we want to compute mathematically is shown by the formula below.

$$X^*_{1:n} = \arg\max_{X_{1:n}} P(X_{1:n}|e_{1:n})$$

Conclusion

Dear Readers,

I sincerely thank you for reading until this point – I hope the information in this book has proved useful and interesting.

This book gave you a brief insight into the world of Markov Models. We discussed the most relevant mathematical equations and high-level concepts, which ultimately define the design and performance of these models. The key topics we focused on included Markov Chains, Hidden Markov Models and Bayesian Networks.

This book is also the third instalment of the **Advanced Data Analytics** series and I hope you have enjoyed it. To view all available titles, please visit my Amazon Author *Page.*

My Most Sincere Gratitude for Reading,

Joshua Chapmann